Cobalt:

blue secret to wealth

Wendy M.J. Michiels

ISBN/EAN: 9789082777932

D/2018/14.286/2

Nur Code : Pockets non-fiction
Nur Code 2 : non-fiction informative/general

www.gaelicvictors.com

Dedicated to Eddy

Strange things are happening in the crust of the earth. Since pre-historic times, there is a lot of activity and messing around underground. But what is expected to be found in this dark, gloomy and unknown world?

Well, just about anything that cannot be cultivated agriculturally, or those materials which cannot be successfully (re-)created artificially in laboratories and factories. Coal, natural gas or limestone are but a few of the materials we can mine, but what fascinates us that much more are the shimmering materials.

Things that sparkle catch our eye, trigger our imagination and curiosity, even our greed in one way or another. And it's not even due to our own human shortcomings that we are so attracted to shiny objects. According to a team of Belgian researchers, our preference for glossy over

What treasures are hidden?

matte
might be
evolutionary-based
and therefore deep-rooted
in every human being.
Glimmering objects, they say, remind
us of our basic need for water.(1) Indeed,
water is one of the most basic needs for us
humans to survive. You can go without
food for weeks. Without water however, you
won't survive for very long. But in spite of
the undeniable necessity of water for

our human survival, maybe we have to dig deeper
(not in the earth's crust just yet, hold on, we'll get
there). Finding our primeval evolutionary link
with water could well lead the way to our obses-
sion with shiny things.

Ever since Darwin, the famous British biol-
ogist, proposed his pioneering but controversial
theory on evolution in his well- known book "On
the Origin of Species" – first published in 1859 –
the research into the origin of man got into a rap.
Science strikes us continuously with proof that
millions of years before the dawn of mankind, the
first life on Earth originated out of the primordial
soup. The molecules swimming in this broth com-
bined into pairs that formed a larger compound.
These building blocks of life would ultimately
write a code of their own, better known as DNA.
As both the base pairs of Adenine and Thymine
danced fiercely with Cytosine and Guanine, newly

formed chemical combinations got the party of life started. Well on its way, the dance resulted further into single-celled organisms, still swimming in the soup of life. Could it be from these earliest of times that our attraction to water stems from?

If all basic building blocks of life, which over time resulted in the existence of you and me, originated in water, can our fascination with this liquid gold – as some investors call it - and possibly derived from it our interest in shiny objects – as not only investors are intrigued by – be simply explained by evolution? How exactly did we evolve from just some letters in a soup bowl to the marvelous creatures we are today? This special interest of us, humans, to have that fundamental question answered will be ongoing until all evidence is revealed and illuminates the true facts. Trying to explain mysterious events and circumstances is just part of our curious human nature.

Centuries ago, before the development of the sciences, our ancestors tried to explain the world around them as best as they could with the little knowledge that they had. In this birth cradle of superstition, myths and sagas - often solely based on observations from a single perspective - clinged strongly into the habits and thoughts of cultures and empires. Originated as oral stories and handed down from generation to generation, these tales are even to this day intertwined with our contemporary views, customs and language.

Not only the gods, strange creatures or heroes take the leading role in these stories. The shiny objects that catch our eye and put in motion the chain of our evolutionary history, causing a commotion in every cell of our body, also form an inexhaustible source for these fantasy tales. Nobody could be amazed by the spell that precious metals can put on us.

As one of the world's most valuable metals, the precious metal gold has not only made a wonderful subject of myths and legends over the years, but also played a significant role in symbolism and folklore.

Due to its attributes , the metal is a symbol for immortality. It will never tarnish, corrode, oxidize or fade in colour. An object of gold, which has been excavated after being hidden underground for hundreds or even thousands of years, will come out in the same condition today as when it was first buried centuries ago. To our medieval ancestors, the attraction of gold was even so great that a whole branch of science was dedicated to it: the alchemy.

As an ancient branch of natural philosophy, the alchemy experimented with elements and materials to produce gold artificially in a labora-

tory from non-precious metals such as lead. The teachings of Aristotle gave the alchemists reason to assume this could in fact be done. And of course, great wealth was promised to anyone who could figure out how to transmute non-precious metal into gold. When the economic interest of this pursue was awakened amongst the popes and rulers of the Middle Ages, alchemy reached its zenith.

A side effect of this quest for gold creation was that the knowledge about extracting metal from ores increased significantly. Yet, the mining of stones and metal was a human activity ever since prehistoric times. The finds from these excavations were generally used to make early tools and weapons. However, discoveries of ancient coloured porcelain in China and coloured glass with its origin in Ancient Persia, Egypt and even Denmark, reveal to us that cobalt compounds for colouring was—even in those ancient times—in high demand. Even today, not only blue, but also red, green and yellow shades can be created with cobalt compounds, for applications such as printing inks and rubber-coloring pigments. Some simple cobalt salts change colour from red to blue, depending on the moist in the air. Take a trip down memory lane, and you will undoubtedly remember the cute but kitschy weather stations or weather dolls, which colour blue in dry air but turn to pink as soon as the air becomes more humid. A fun but harmless use of cobalt. The mining, however, of cobalt was not at all without danger in earlier times. The underground labour in the rocks was hostile and full of dangers.

Kobalt is found in metal ores often along copper, nickel or iron? It was a very unpopular and even frightening material long before its discovery as a chemical element around 1730 by Swedish scientist Georg Brandt. Because of its silvery appearance, cobalt was often mistaken for other desired metals by the early miners. When using heat underground, toxic arsenic fumes were formed, which of course posed major problems such as serious illnesses, sometimes even with deadly consequences. Medieval miners, not understanding the composition and workings of the materials around them, blamed the evil mountain spirits for their misfortune, a belief that was very common in those days. These mischievous gnome-dwarfs were named after the Greek word Kobolos, which translates as rascal or villain. Kobolds are believed to live underground, mainly engaged in digging up the shiny metals that are hidden like treasures in the earth's crust.

Today, the Democratic Republic of Congo is one of the world's leading sources for mining cobalt. The African country mines more than half of the world's supply. After delving, the raw cobalt ore gets industrially refined, mostly for use in the battery industry.

Due to the increasing demand for batteries for electric cars, amongst other battery-powered appliances, the price of cobalt on the world market is on the rise. Reforms in the mining laws of Congo relating to the strategic metals, of which cobalt undoubtedly covers a huge part, are currently the subject of debate. Rumor has it that even a world-renowned tech giant is in the middle of securing its cobalt supplies.

Maybe it's time for us too to dig a little deeper into this fascinating commodity. Perhaps the future colour of wealth might not be dollar-green but a beautiful deep cobalt blue.

Wendy M.J. Michiels

Good to know...

In 1957 the Danish discovered the stabilizing effect of cobalt compounds on the foam head of beer. Although cobalt is an essential part of our body as part of vitamin B12, too much cobalt is very dangerous and the experiment was banned soon after. As it turned out, cobalt was even worse for the liver than alcohol.

Cheers!

Wendy M.J. Michiels

References:

1. *All That Glistens: II. The Effects of Reflective Surface Finishes on the Mouthing Activity of Infants and Toddlers*

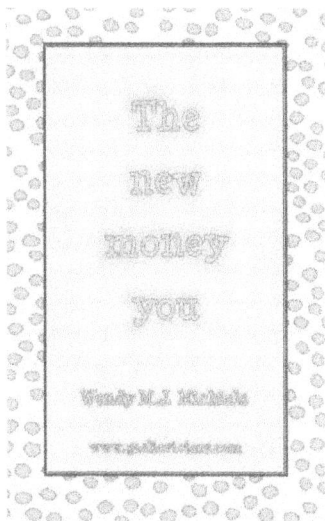

The new money you

Coaching and publishing your creativity

www.gaelicvictors.com

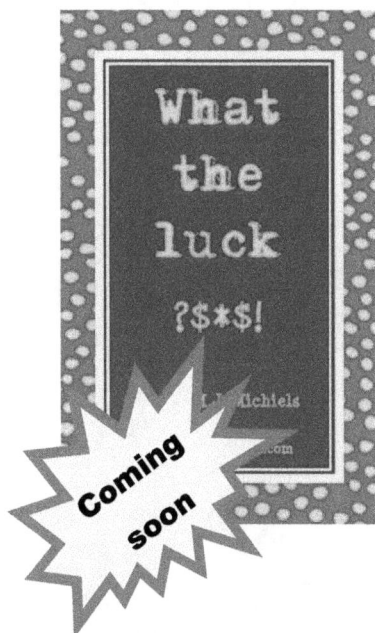

What the luck ?$*$!

Coaching and publishing your creativity

www.gaelicvictors.com

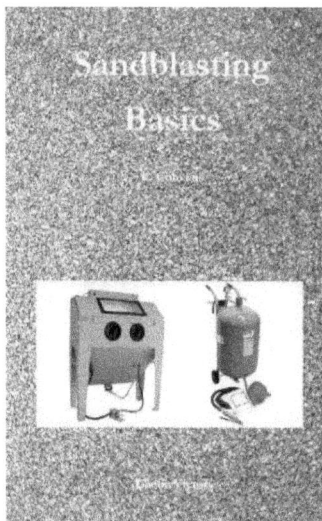

Sandblasting Basics

Coaching and publishing your creativity

www.gaelicvictors.com